AUSTRALIAN TROPICAL
Butterflies

by
Peter Valentine

Photography
Clifford & Dawn Frith

Published in Australia
by Tropical Australia Graphics
'Prionodura' Paluma via Townsville,
Queensland 4816. (077) 708523

National Library of Australia
Card Number ISBN 0 9589942 5 0

First printed 1988

Other books in this series:
Australian Tropical Rainforest Life
Australian Tropical Birds
Australian Tropical Reptiles and Frogs
Australian Tropical Reef Life

Preface

In 1975 I came to northern Queensland to study tropical
rainforests and one of the first things I noticed was the
large number of butterflies wherever I went. Subsequently,
found the butterflies so fascinating that I began to take a
serious interest in them as subjects for study. In this I was
greatly assisted by the excellent text on Australian
butterflies by Common and Waterhouse (see page 69), and
by continuing encouragement and practical help from a
great many people.

When Cliff and Dawn Frith suggested a butterfly volume
for their successful book series I was delighted to assist. It
would be marvellous to see some of the beautiful butterflies
of tropical Australia presented in colour photographs and
was decided that only species which could be
photographed alive in a natural setting would be included.
was also decided to present the selection within a
framework of habitats and six distinct environments were
identified and species allocated to each depending on the
most likely country in which each butterfly could be found.
In this slim volume we present 129 colour photographs
of 118 species of Australian tropical butterflies. Included
are 30 of the 37 species collected by Banks and Solander
during Captain Cook's voyage up the eastern coast in 1770.
describing each butterfly the binomial species name is
used and subspecies are not differentiated in maps or text.
scientific names given by Common and Waterhouse are
followed but new distribution, behaviour, and life history
information has been incorporated where possible and
where recent taxonomic changes have been suggested
these are indicated in the text or in brackets. Because most
people using this book will be more comfortable with
English common names these are used to head species
accounts. At the end of each text the family to which the
species belongs is given. Maps show the approximate
known range but it should not be assumed the species
occurs everywhere in areas indicated.
Of the butterflies illustrated 49 are exclusively tropical
and a further 50 are mainly tropical but also occur south
of the Tropic of Capricorn. The remaining 19 species are
widely distributed in temperate Australia but also occur in
least part of the tropical zone. It is hoped that this book
will enhance the reader's appreciation of butterflies and
the need to protect their habitats.
In acknowledging the contributions of many people to
my knowledge of butterflies I especially thank Steve
Johnson. It has been of particular value to have enjoyed his
excellent company in the field and his generous sharing of
firsthand knowledge of the fauna. We have undertaken
much together, have shared many publications of our
discoveries and my debt to him is great. For company in
the field and other assistance I am also grateful to Arnie
Johnson, David Lane, Max Moulds, Geoff Monteith, Chris
Hill, John Kerr, Don Sands, Steve Brown, Russell Mayo,
Ann Young, Naomi Pierce, Peter Samson, Gary
Sankowsky, Peter Wilson, Graham Wood and Trevor
Lambkin. In completing this book the cooperation
commenced in the field, between author and
photographers, was further extended. Much of the text
benefited from wise advice and practical assistance from
Cliff and Dawn Frith and I thank them for this and for the
friendship cemented in the process. Without the
understanding and support of my wife, Valerie and our
children, Jody, Leonie and Kate, I would have found the
challenge of discovery impossible to accept. My obsession
with butterflies has cost them much of my time and for that
I hope they forgive me.

Many people kindly helped the photographers with the
location of butterflies, hospitality, companionship in the
field, and advice. We sincerely thank Mick and Vivien
Atkinson, Michael and Madelaine Beach, Len Cook, Bill
and Wendy Cooper, Ted Fenner, Nicola Frith, Andree
Griffin, Brian and Helen King, Barry and Shelley Lyon,
Roy and Margaret Mackay and family, Jeff and Jo McClure,
Arthur and Margaret Thorsborne, Bill Travers, Peter and
Valerie Valentine, and John and Janelle Young.

We gratefully acknowledge the help of Paul Wright and
Gary Sankowsky of Kuranda Butterfly Sanctuary, and
Michael and Jaki Turner of Daintree Butterfly Farm in
providing access to butterflies in their care. Peter Samson,
Peter Valentine and Australasian Nature Transparencies
picture library provided pictures of butterflies we were
unable to take for this book; their pictures are credited
beneath relevant illustrations.

It is a pleasure to express our gratitude to Peter
Valentine for writing the text for Australian Tropical
Butterflies. His extensive experience with and knowledge
of our tropical butterflies has enabled us to provide a book
as accurate and interesting as it is colourful. Moreover,
working with Peter has been a great pleasure, resulting not
only in this book but also in a close and valued friendship
with him and his family.

Our photographs were taken exclusively with
OLYMPUS cameras, flash systems, accessories and
ZUIKO lenses. We use the remarkable Olympus OM–4
camera body and a wide range of lenses. Of particular
valuable application for this book were the innovative and
advanced macro photographic systems developed by the
Olympus Optical Company; specifically the 65–116 mm.
telescopic auto-tube with 135 and 80 mm. macro lenses,
the 50 mm. F3.5 macro lens, the auto extension tubes, and
the T28 macro twin flash system with ' power control 1
and six volt power pack 2. All photographs were taken on
Kodachrome 64 KR135 film. Our tropical travel was by one
ten long wheel base Land Rover.

Peter Valentine,
Geography Department,
James Cook University.

Clifford and Dawn Frith,
Paluma, Tropical North
Queensland.

1

Introduction

Butterflies and the Tropics

Butterflies occur throughout the world apart from the polar regions but reach their greatest diversity in the tropics. Here butterflies are free from extremes of temperature and the threat of aridity, and reproduce at quite remarkable rates. Some tropical species may grow from egg to adult in three weeks. In contrast at least one species in southern Australia takes two years to complete the same process.

The most important resource for butterflies is their larval food and for most that means foliage, flowers or fruits of living plants. The abundance of this resource in the tropics helps account for the numbers of butterflies evident in northern Queensland. Of all 385 known Australian species about 330 (86%) occur in Queensland, considerably more than in any other state. There are only 39 in Tasmania and 64 in South Australia, reflecting the two factors of cold and aridity respectively. Within the tropics occur 300 species of which 136 are exclusively tropical.

A commonly asked question is how do butterflies differ from moths. An easy distinction was once made — butterflies had clubbed antennae and moths did not. More recently it has been recognized that many moths have clubbed antennae. Moths and butterflies belong to a large group of insects known as Lepidoptera ("scale wing"). All butterflies have clubbed antennae and do not have devices to lock their wings together in flight (except the male Regent Skipper — page 12). Day-flying moths have devices to lock wings together in flight. The technical distinction between moths and butterflies is thus relatively simple in all but a few cases. In tropical Australia, most colourful day-flying Lepidoptera seen will be butterflies.

The approximately 17,200 butterfly species on earth are divided into a number of families; groups sharing many characteristics and which appear to have a common evolutionary history. In this book six families are recognized following the text by Common and Waterhouse (see page 69). The Hesperiidae is a group of mainly tropical species known collectively as skippers, but includes considerable diversity of form. Perhaps the most impressive family is the Papilionidae which includes the large colourful swallowtails and birdwings. Almost entirely tropical in Australia is the family of whites and yellows, the Pieridae.

The largest Australian family is Lycaenidae (blues) with 136 species (6000 world wide) while the smallest is Libytheidae, with only 1 species (page 35). Finally the family Nymphalidae (6000 world wide), is divided by some into many separate families. Others classify these differences at the subfamily level. True nymphs (Nymphalinae) are primarily butterflies of tropical rainforest, large and colourful; but the family also includes milkweed butterflies (Danainae), the mainly temperate browns (Satyrinae), and seven or eight small subfamilies.

Basic Butterfly Biology and Behaviour

A butterfly spends much, if not most, of its life in immature stages — as an egg, larva (caterpillar) or pupa (chrysalis). The adult form is the most familiar stage, flitting from plant to plant in gardens. This final stage i brief although some species survive for many months i "overwintering" populations. Many smaller butterflies for only a week or two. Butterfly larvae are fussy eaters each species or group of species being restricted to a fe plant species. A few butterfly larvae are carnivores, preying on other insects (see Moth Butterfly, and Large Ant-blue, pages 42 & 60).

The pupal stage of a butterfly is a period of major change — the transformation of caterpillar into winged adult. This metamorphosis is a wonder of nature. Pupa are usually attached to the food plant, or to some other convenient object, by a silk strand spun by the larva be it pupates. After a period of one to three weeks (in the tropics) the adult butterfly emerges. It will usually hang from a twig and slowly inflate its wings, allowing them harden and dry before flight. Then follows the most ac period of the butterfly's life and its primary function is reproduction which may require dispersal to find larva foodplants, sometimes a short distance but at other tim involving longer migration in large numbers.

In some species finding a mate is not a problem as m wait near the pupae, often copulating with freshly eme females before their wings have dried. Others have less chance to find a mate and these adopt a strategy know "hilltopping". Males congregate on hill summits and establish territory, defending perches and adjacent ter from other males. Eventually a female will arrive and b quickly located by a male. Normally females do not ret to the hilltop after copulation, but males stay there for remainder of their lives. Most adult butterflies take nec from flowers and a few imbibe juices from rotting fruit Many take water from soaks. Some, however, survive entirely upon accumulated fat. After mating, females s out food plants and lay eggs, singly in some species, in large clusters in others. Many plants are utilized by butterfly larvae, from grasses to herbs, shrubs, orchids trees, including mangroves and mistletoes. In the fami Lycaenidae (blues) many species have special relations with ants and females will only lay eggs when the corr plant and ant occur together. Ants attend larvae, gainin sustenance from liquids produced by the larvae and conferring a degree of protection to them from parasite and predators. The larvae and pupae of some ant-atten species produce sounds! Pupae in particular have an astonishing ability and when softly stroked, or even blo upon, will produce a loud "brrr . . . brrr' sound. If seve are gathered together the sound is quite audible. I have noticed green ants react to this sound and it is possible it may alert the ants to disturbance.

Upland Rainforest

Ulysses Butterflies mating

Ulysses Butterfly
Papilio ulysses

This beautiful butterfly (**right**) is a symbol of tourism throughout northern Queensland. The brilliant blue of the upper wings is a spectacular sight as these very large insects fly within rainforest openings. They are commonly seen at lookouts soaring above the canopy then swooping to the ground with the blue flashing on and off as the wings open and close. The underside is cryptic when the butterfly (**page 3**) is at rest.

Although found commonly from Mackay north they appear more prevalent in upland rainforest and an alternative common name is Mountain Blue. The Queensland Government has passed legislation protecting this species although it is in no sense endangered as a considerable area of its habitat is protected in National Parks. Its numbers have probably increased in urban areas recently due to plantings of its larval food plant (*Euodia elleryana*). Papilionidae

Nigidius Jezabel
Delias ennia

The jezabels are an attractive group of butterflies which are white above but colourful beneath. Found from India to South-east Asia they are particularly common in Papua New Guinea. The larvae all feed on mistletoes or related plants and in Australia six of the eight species occur primarily in the tropics. This species (**opposite**) is represented in Australia by two subspecies and is confined to rainforest. It is a common butterfly at Paluma and at Kuranda. The photograph shows a male freshly emerged from its pupa clinging to its larval food plant, the Golden Mistletoe. Pieridae

Union Jack
Delias mysis

This delightful species (**page ii**) is a commonly seen butterfly of the tropical rainforests in northern Queensland. The upper side is almost entirely white in contrast with the brilliant red and yellow of the under side. Females have more extensive areas of black. Eggs are laid in clusters on the leaves of the mistletoe food plant and the green larvae feed together in rows until they are ready to pupate. The pupa is suspended by a silk girdle and several may be seen together on the underside of a leaf. Pieridae

Helena Brown
Tisiphone helena

Most members of the Satyrinae subfamily of butterflies are temperate and all the close relatives of this species occur in southern Australia. It is therefore not surprising that the Helena Brown (**right**) is mainly found in cooler upland rainforests. Despite the more uniform climate of the tropics adult Helena Browns have retained a strong seasonality and are on the wing in summer only. It is common to see them flying in and out of the rainforest amongst the clumps of sword-grass upon which the larvae feed. Nymphalidae

Australian Hedge Blue

Celastrina tenella

This small species is best characterized by the word elusive. It is rarely encountered and is confined to rainforest in the Kuranda area although occasionally seen at Danbulla State Forest and at Crystal Cascades near Cairns. The male (**right**) is pale blue above and the female mainly black and white. Nothing is known of its life history. Other species of *Celastrina* occur in Papua New Guinea and throughout South-east Asia to India. At Kuranda the butterfly is mainly seen in winter although the photograph was taken in February at Clohesy River. Lycaenidae

Blue Triangle

Graphium sarpedon

Widely distributed from Asia to Australia this robust flyer (**opposite above**) is commonly seen swooping low over flowers, or sipping at nectar along the edge of rainforest. Males will frequently fly right up to people wearing blue clothing and even alight momentarily. The upper wing produces a flash of blue as it flies. The bluish-green humped larvae usually feed on the leaves of laurels. Triangle butterflies are powerful and rapid flying species with distinctive and incessant quivering wingbeats. Papilionidae

White Lineblue

Nacaduba kurava

Walking along the edge of upland rainforest you will almost certainly encounter a few adults of this species (**right**) flying slowly a few metres or less above the ground. These are likely to be females. Males frequently take up territory on the upper surface of a leaf in a sunny glade where they keep watch for intruders which are rapidly seen off. The extensive white patch on the underside of the wings makes it easy to identify this small butterfly. At Paluma females lay their eggs on the buds of a low spreading shrub, *Maesa dependens*, and a careful search might reveal the tiny green larvae. Lycaenidae

Macleay's Swallowtail

Graphium macleayanum

The rich green colour of the underside of this species (**opposite below**) is distinctive and contrasts with the white and black upper surface. In the tropics Macleay's Swallowtails are commonly seen at blossom in upland rainforest areas, especially sipping nectar from *Lantana* flowers. Remarkably, this is also one of the few species which can be seen flying in the Australian Alps where it is on the wing for a brief period each summer. In northern Queensland it flies throughout the year. One subspecies occurs on Lord Howe Island and another in the highlands of Papua New Guinea. It is the only member of the Papilionidae family of butterflies to occur in Tasmania! Despite this temperate affinity it is a very common butterfly of upland rainforest in tropical Queensland. Papilionidae

 ## Nysa Jezabel
Delias nysa

Although the Nysa Jezabel (**left**) is widely distributed in southeastern Australia, in northern Queensland it is mainly seen in upland areas. Like the Grey Albatross and Common Jezabel (below) this is one of the 37 species of butterflies collected in Australia by the botanist Banks in 1770 and was most likely taken while Captain Cook's ship the Endeavour was being repaired at what is now Cooktown. One of the small jezabels, it is distinguished by the extent of black below and the yellow markings. The upperwing is white with narrow black margins in the male and broader margins in the female. Pieridae

 ## Grey Albatross
Appias melania

Found from Cooktown to Townsville this species (**opposite above**) is often abundant at Paluma where it may be seen feeding on flowers along the edge of the rainforest. At times hundreds migrate and they have been seen in large numbers on the summit of Mt Stuart in Townsville. The upper surface is mainly greyish-white with a few black markings. The Grey Albatross is only found in Australia and is a close relative of the jezabels. Pieridae

 ## Peacock Jewel
Hypochrysops pythias

How unfortunate that this aptly named gem (**left**) is unknown to most visitors to the upland rainforests. Confined to the area from Cape Tribulation south to Bluewater State Forest near Townsville, it is a not uncommon species within its range. The males, a rich purple above, are aggressive guardians of the canopy and rarely descend to the lower forest. Along the edge of forest tracks they may be seen engaged in aerial battle — small dark specks flying at great speed. The females are more down to earth creatures, searching along disturbed areas for *Commersonia bartramia* trees on which to lay their eggs. Larvae shelter under leaves and are well camouflaged. With good luck females can be observed in the act of egglaying and the brilliance of their colour appreciated. Lycaenidae

 ## Common Jezabel
Delias nigrina

In northern Queensland this species (**opposite below**) may be found commonly in upland rainforests but it also occurs in coastal areas of NSW and southern Queensland. The beautiful red marking on the underside of the hind wing is distinctive but the dark background colour underneath can make the butterfly difficult to see when at rest. In flight the bright white upper wing of the male makes it conspicuous although the female is much less so. Most of the jezabels are on the wing throughout winter months in northern Queensland and they are usually a dominant feature of a visit to upland rainforest. Pieridae

Purple Brown-eye
Chaetocneme porphyropis

Found from Cape Tribulation to Paluma this species was poorly known until recently. One reason for this is that the adults fly mainly at dusk and are rarely encountered. The female lays her eggs singly on the leaves of rainforest laurels, favouring *Neolitsea dealbata* and *Litsea leefeana*. The young larva constructs a shelter by cutting a small piece of leaf and bending it over to form a tent, secured by silk. As it grows it constructs successively larger shelters until it pupates in the final one. The rather tatty adult (**right**) was photographed in Paluma, the first record this far south. Hesperiidae

Black Skipper
Toxidia melania

This is another species with a very limited distribution, confined to the rainforests from Mossman to Townsville. Another similar species occurs on Cape York from Iron Range north (*Toxidia inornata*). The Black Skipper (**below**) is a summer butterfly being found on the wing from about January to April. In common with other skippers adults are very rapid flyers and males are aggressive in defending their vantage points. At Bluewater State Forest, near Townsville, they can be very common at times where they fly in areas with swards of wire grass in more open patches of rainforest. The larvae feed on the wire grass. Hesperiidae

Black and White Swift
Sabera caesina

What a well named butterfly! The adult (**left**) is crisply marked in black and white with beautiful white antennae. As they flash in to a patch of blossom or to a sunny perch the term swift is seen as perfectly apt. Confined to the rainforests of northeastern Queensland they are commonly encountered in disturbed areas. The larval food plant is the well known lawyer vine, a scrambling palm (*Calamus* species) which does well in canopy gaps produced by tree falls.

When resting in a sunny patch the adult will open its hind wings to better capture the warmth of the sun. They are enthusiastic feeders at nectar and may spend many seconds drawing up this high energy food before zipping off to another flower. Hesperiidae

Common Red-eye
Chaetocneme beata

Although found as far south as Sydney, this species (**below**) seems to be an upland rainforest butterfly in northern Queensland where it occurs north to Cairns. At Paluma the larvae occur on *Neolitsea dealbata*, a laurel which has leaves with a beautiful greyish-blue underside colour. As is the case with all the red-eyes the larvae build tent-like shelters. In northern Queensland the Common Red-eye occurs along the edges of rainforest or in wet sclerophyll forest. Hesperiidae

Miskin's Jewel
Hypochrysops miskini

In Australia there are eighteen species known of this group of exquisite butterflies and several are illustrated in this volume (see index). Miskin's Jewel has an interesting distribution, occuring in rainforests from Kuranda to Townsville with a large gap until the rainforests of southeastern Queensland. The photograph (**right**) shows a female, which is considerably paler than the male. This species is quite abundant in the upland rainforest but males are not often seen as they assemble on the canopy of the forest where they engage in spectacular aerial battles over perches. The females, once mating has occurred, remain close to the forest floor where they seek out the plants upon which they lay their eggs. Originally it was thought this species used only the vine *Smilax australis* but it is now known the larvae feed on many different plants. What is more important is the presence of a particular species of ant. Miskin's jewel has a special relationship with a rainforest ant, *Iridomyrmex gilberti*, which attends the larvae and pupae. The ants "milk" the butterfly caterpillars from special glands which yield a nutritious liquid. In return they give protection from predators and parasites. Lycaenidae

Regent Skipper
Euschemon rafflesia

The magnificent adult of this species is shown soon after emerging from its pupa (**page i**) the empty skin of which is visible above it. The equally spectacular larva (**right**) feeds on a common understory species of upland rainforest. Another subspecies of this butterfly occurs in southern Queensland and in northeastern NSW where some concern has been expressed for its survival. As is inevitably the case in Australia, butterfly survival is totally dependent on habitat protection. Apart from its great beauty, the Regent Skipper is of special interest to science for it is perhaps the most primitive of the world's butterflies. One of the distinguishing features between butterflies and moths is that the latter possess an organ (called a frenulum) which locks the fore and hind wing together during flight. Butterflies do not have these. The exception to this rule is the male Regent Skipper (females are more advanced as they have lost theirs!).

Adult Regent Skippers usually fly within the rainforest but are frequently seen taking nectar from blossoms. Females may be encountered diligently searching for the larval food plant which they usually fly around several times before selecting a site at which to lay their egg. When resting, especially in sunlight, the adults open their wings almost flat. They are often seen peering out from the edge of a leaf, a stance which displays vividly the scarlet head and underbody. Hesperiidae

Hamadryad
Tellervo zoilus

It is now well understood that Australia once formed part of Gondwanaland — a continent linking Africa, South America, India and Australia. Subsequent separation has produced distinctly different flora and fauna in these countries but there are, however, just a few species of plants and animals which help demonstrate that ancient link. The Hamadryad (**left**) is one of these. Sole local member of the subfamily Ithomiinae this single species is very different from other Australian butterflies but has close relatives across the Pacific in southern and central America. The yellow eyes are distinctive. This attractive butterfly is frequently seen slowly fluttering and gliding in and out of rainforest openings. Its larval food plant is a large-leaved vine (*Parsonsia* species) which commonly climbs on trees along track edges. The larva is colourful with a pair of black fleshy filaments rising from bright yellow bases immediately behind the head. The body is narrowly banded black and white with a pair of yellow patches at the rear.

A second subspecies occurs on Cape York with others in the islands immediately north of Australia. The adult Hamadryad is similar in appearance to the Black and White Aeroplane, (see page 19), a species to which it is unrelated. This may represent an example of mimicry by the aeroplane to gain benefit from the toxicity of the Hamadryad. Nymphalidae

Spotted Skipper
Hesperilla ornata

The sword-grass which occurs in tropical rainforest areas is the larval food of this lovely skipper (**left**). In northern Queensland it is commonest in the upland rainforest although it does occur on the coast (for example in Conway National Park at Shute Harbour and in the Hinchinbrook Channel National Park south of Cardwell). The genus *Hesperilla* is an entirely Australian group of butterflies which are mostly southern but with five species in the tropics. The Spotted Skipper is the most brightly coloured of these. The adults are black with orange markings above and beautifully spotted black and white below as the photograph shows. Males are territorial and even females display the rapid direct flight typical of skippers. Larvae feed upon the blades of sword-grass, constructing tubular shelters in which they hide during the day. They emerge at night to feed leaving distinct triangular notches in the leaves. The larva pupates within the shelter. Hesperiidae

oto: P. Valentine

Female Birdwing Butter

Australian Lurcher
Yoma sabina

This large and distinctive butterfly (**page iii**) is typical of the tropical nymphalids. It seems to prefer a mix of lowland rainforest and swampy areas and is quite common in Cooktown and on Thursday Island. The adults are usually nervous and do not allow a close approach. Although the upper surface is richly coloured the underside is quite cryptic and adults can be difficult to locate once they have settled on a tree trunk or branch. Females frequent swampy areas where larval food plants occur, (*Ruellia* species — a herbaceous shrub). Nymphalidae

Birdwing Butterfly
Ornithoptera priamus

This magnificent creature, also referred to as Cairns Birdwing, is Australia's largest butterfly and the female (**opposite**) is shown drying her wings after emerging from a pupa. The male is richly coloured in green, gold and black (**cover upper left**). There can be few sights as spectacular as a soaring birdwing gliding amongst the canopy or sweeping down to investigate a flower. Writing in 1856 the great naturalist Alfred Wallace described his joy at seeing a birdwing during his sojourn in the Malay Archipelago. To him *Ornithoptera* were the "largest the most perfect and most beautiful of butterflies". Many people today would share that view and in Papua New Guinea birdwings are carefully farmed.

Once at Iron Range, on Cape York peninsula, I saw several Birdwing Butterflies feeding on the massive yellow blossoms of the Golden Bouquet tree (*Deplanchea tetraphylla*). Sharing the nectar was a Graceful Honeyeater, completely dwarfed by the butterflies! Males sometimes set up territory near the larval food plant, a vigorous large-leaved vine *Aristolochia tagala*, and in the early morning patrol around its base in search of freshly emerging females. These may be mated before their wings are dry and it is not unusual to see two or three males trying to copulate with a single female. By planting vines in their yards many people have established permanent populations of these spectacular butterflies. Papilionidae

Cruiser
Vindula arsinoe

There is sometimes a surprising difference in appearance between males and females of the same species, as is seen with the Birdwing Butterfly. In the Cruiser this sexual dimorphism is quite dramatic. The male (**left above**) is orange above and usually somewhat smaller than the female (**left**) which is mainly brown above. Both fly strongly but may be readily seen feeding at nectar. The larval food plant is a native passion vine, *Adenia heterophylla*, and the female lays the eggs in clusters. The pupa is heavily flanged and dull coloured, looking very like an old dead and twisted leaf. At times the adults are very common and are usually encountered feeding on *Lantana* flowers at the edge of roads. Nymphalidae

15

Black and White Tit
Hypolycaena danis

It is difficult to imagine such a lovely insect as a pest but the Black and White Tit is well known as such to orchid growers. The adult butterfly (**right**) lays its eggs on the buds of epiphytic orchids and the larvae feed on the flowers. In the Cairns area they create much damage for commercial growers who need to keep their plants well sealed in shade houses. The butterfly is known from Papua New Guinea and Cape York to Wallaman Falls, near Ingham. They are rarely seen in any numbers in the rainforest, partly due to the dispersal of their larval food plant and its location in the canopy. Lycaenidae

Ambrax Butterfly
Papilio ambrax

At a distance this species resembles the better known Orchard Butterfly, but unlike the latter it is confined to rainforest. The female (**opposite above**) is more striking than the male and is usually seen flying along the edge of rainforests or in a clearing. The species is also known from Papua New Guinea so it is surprising that it has not been found north of Cape Tribulation. The Ambrax Butterfly lays its eggs on plants in the family Rutaceae including the introduced citrus trees. Papilionidae

Orange Aeroplane
Pantoporia consimilis

Almost any patch of rainforest in northern Queensland will have a small population of these attractive creatures (**right**). Their gliding flight is distinctive and displays their orange and chocolate colours beautifully. Individuals rest on the upper surface of a leaf with wings flattened to catch the sun. Suddenly one might fly rapidly out in the direction of an intruder, or do so for no apparent reason, and then cease beating its wings to glide around in an arc or return to a vantage point. Nymphalidae

Capaneus Butterfly
Papilio fuscus

Although confined primarily to lowland rainforest this species (**opposite below**) also occurs in deciduous vine thickets, and in the wet tropics will lay its eggs on cultivated citrus trees. The pupa has the ability to stay dormant for a very long period, sometimes longer than a year. In Townsville it can be a very common species in good wet seasons, but from about 1982 until 1987 it did not appear in any numbers, coinciding with a run of very dry seasons. In summer this is a remarkably common species in the vine thickets of Cape York and larvae may be seen on several of the Rutaceae plants common to those communities. The pupa is bright green. A very close relative, the Canopus Butterfly (see page 35), is confined to the Northern Territory. Papilionidae

Photo: P. Samso

Green-spotted Triangle
Graphium agamemnon

Along a rainforest track is the best place to see one of these brilliant butterflies. Their most distinctive characteristic is the constant ''nervous' flutter which they maintain even while feeding at nectar. Swooping down from the canopy an adult will pause momentarily at a flower, hardly long enough to insert its proboscis, before rushing on to the next. It is rare that they slow down long enough for the photographer to capture the beauty of the upper wings (**opposite above**). The underside has a much duller appearance although it is enlivened by two or three red spots.

Occasionally Green-spotted Triangles will fly in urban areas where females use the introduced soursop tree on which to lay eggs. The larva is chocolate brown but with a white central saddle distinguishing it from the Pale Green Triangle larva which also feeds on soursop (see page 67). Papilionidae

Australian Leafwing
Doleschallia bisaltide

This large tropical nymph provides a superb example of cryptic camouflage. When resting in sunlight the adults frequently open their wings and display the rich orange and brown upper surface (**left above**). The adult frequents the forest floor and when in shade always rests with wings tightly closed above its head. In this position (**left below**) it looks very like a leaf. The hind wing has a projection which mimics the leaf petiole and the thin line running across from forewing to hindwing mimics the midrib of a leaf. Even the colour is a perfect match for a dry leaf. I have frequently failed to locate adults which I watched land because they have blended with the surrounds so well.

One reason the females fly close to the ground is to search for their larval food plant. This is a small herb with lovely white flowers and purple underside leaves (*Pseuderanthemum variabile*). The egg is laid on the buds of the flower stem and the larva feeds first on buds and later on the leaves. The mature larva leaves the plant and pupates under a rock overhang or under a leaf on a nearby tree. Nymphalidae

Black and White Aeroplane
Neptis praslini

Like the Hamadryad (see page 13) which it resembles, the Black and White Aeroplane has yellow eyes (**opposite below**). But this resemblance is not due to similar genes as they are not closely related. No doubt the non-toxic but possibly palatable aeroplane achieves a benefit from this resemblance to the toxic Hamadryad. Careful attention to the flight pattern reveals some difference as the aeroplane appears to have a much stronger and more direct flight. The larvae resemble the larvae of the other Australian aeroplane butterflies and also feed on plants in the family Fabaceae — usually rainforest climbers. The species is known from Cape York to Paluma. Nymphalidae

19

Dark Ciliate Blue
Anthene seltuttus

This small blue butterfly, (**right**) has a close relationship with the Green Tree Ant (*Oecophylla smaragdina*). The female lays its cluster of 30 — 40 eggs only on a plant infested with green ants. Once the eggs hatch the tiny larvae feed on the buds or fresh foliage always attended by the ants. The ants lick a fluid from special glands and careful examination of a large larva will reveal the periodic protrusion of an organ which the ants lick. Usually the large larvae gather together to pupate, typically on a small twig or branch of the food plant. Even the pupae are attended by ants. The reward for these butterflies providing food for the ants is twofold. In the first place the ants do not devour the larvae despite the fact that they do eat many other caterpillars. In the lowland rainforests and vine thickets of tropical Australia that is a major benefit. The second advantage of close attendance by ants is the reduction of parasitism. It is a very brave or foolhardy wasp or parasitic fly which comes too close to a caterpillar surrounded by green ants!

The adult male is purple above and the female a dull dark grey with a central patch of blue. In recent years this species has become common in parks and gardens where it flies around the lovely Golden Shower Tree (*Cassia fistula*) — but only if the tree has many green ants on it. Lycaenidae

Common Tit
Hypolycaena phorbas

As with the other three species on these pages, the Common Tit (**right**), has had to adapt to life in an environment with many Green Tree Ants. The solution in each case has been to live with the ants and larvae of the Common Tit do not occur without a large number of green ants in attendance. Unlike the Dark Ciliate Blue (above), however, the larvae are solitary and prefer to remain in a shelter when not feeding. One of the interesting things about butterfly larvae adapted to close attendance by ants is that they frequently feed on many different types of plants. Common Tit larvae are able to feed on at least 15 different plants in northern Queensland including 2 mangroves.

The adult male is black above with a dark blue central area while the female is black and white. The underside is similar in each. One notable feature is the presence of tails and false eyes on the underside. These confuse predators such as lizards and birds into biting the expendable end rather than the head. It is not uncommon to find butterflies with bite marks on their hindwing where the tail used to be! Males of this species are quite aggressive and take up a vantage point in a tree from which they can furiously attack any intruders. Lycaenidae

Tailed Green-banded Blue
Danis cyanea

Wherever the huge pods of the Matchbox Bean hang from the forest canopy the Tailed Green-banded Blue usually occurs. This lovely butterfly (**left**) lays its eggs on the fresh shoots of that vine and the larvae, pale green wrinkled caterpillars, are attended by ants. The adults seem to prefer flying under the canopy and at times they are abundant, especially in the gallery rainforest along the lower reaches of streams and rivers. Unfortunately their preferred habitat is usually shared with vast numbers of mosquitoes and despite the attraction of the butterflies only the hardy human lingers long.

I once grew a Matchbox Bean in my garden in Townsville and was delighted when it was used by this species. At the time green ants had taken over the vine but they did not harm the butterfly larvae. They did not appear to attend them as closely as they do some other species of blue butterflies. The adult male is a brilliant blue above and the female black and white. Lycaenidae

Common Oakblue
Arhopala micale

The dull and cryptic underside of this species (**left**) is in sharp contrast with the brilliant blue of its upper wings. Unfortunately it rarely opens its wings except in flight when the alternating flash of blue is quite arresting. The tactic of flying rapidly into thick foliage and perching with wings closed is a very successful disappearing trick. When oakblues are flying around a tree a shake will startle a cloud of them where only two or three might have been expected. They quickly return to hiding places. The females lay their eggs singly on the leaves of at least a dozen different species of trees and shrubs, but only in the presence of Green Tree Ants. The larvae hide when they are not feeding, in curled leaf shelters constructed by the green ants, or sometimes in the entrance to the green ants brood nests. A large oakblue larva may have twenty or thirty green ants in attendance and the ants become quickly aroused and very aggressive if disturbed. Despite this quite a few oakblue larvae are parasitized by a fly, perhaps by the adult fly laying its egg on the leaf of the plant which is then ingested by the butterfly larva. Lycaenidae

Blue Moonbeam
Philiris nitens

On the wing this small butterfly is a silver flash as it flies rapidly from tree to tree. The upperwing is bluish-purple in the male and black and white in the female but below is brilliant white in each with a tiny black dot (**right**). Six species of moonbeams are known from Australia and others may yet be discovered. All except one is confined to the tropics. The Blue Moonbeam occurs as two subspecies in Australia from Cape York to the McIlwraith Range and from Cooktown to Townsville. In the Bluewater State Forest near Townsville larvae are found on *Macaranga* species, common shrubs of rainforest areas. At the South Johnstone area it flies with the Purple Moonbeam, *Philiris fulgens*, both species engaging in aerial battles with others of their kind. Lycaenidae

Australian Vagrant
Vagrans egista

An attractive nymph, the vagrant (**opposite above**) is a butterfly of the rainforest. Their bright orange wings bring flashes of colour against the green forest as they fly along a track or in an opening. Sometimes they will patrol a creek and they are common at times near Cairns at Crystal Cascades. By the middle of the day most males are flying high where they stake out an overlooking perch and venture forth in short flights before returning. Females seek out the rainforest plant *Xylosma ovatum* upon which they lay their eggs. Larvae feed upon fresh juvenile leaves and older larvae have a yellow stripe running the full length of their spiny body. Nymphalidae

Broad-margined Grass Yellow
Eurema candida

Confined to the far north of Cape York peninsula, south to the McIlwraith Range, this is a denizen of rainforest. The male (**right**) is rich yellow with very broad black margins but the female has white above rather than yellow. At Iron Range they are common along the tracks where they fly gently from one roadside flower to another, sometimes disappearing into the forest, at other times content to patrol the verge. Pieridae

Small Oakblue
Arhopala wildei

The most elusive of all the oakblues, the life history of this species is still unknown. Confined to lowland rainforest its habits appear quite puzzling. Adults rest within thick foliage and are reluctant to fly. When disturbed they rapidly return to a patch of vine or other dense cover. Females and males both occur together but despite considerable effort of observation no clues have been uncovered as to the larval food plant. The underside of this species (**opposite below**) is very different from that of other oakblues (see page 21) and the upperside is also distinct, blue in the male but white in the female. It is often in the company of moonbeams. Lycaenidae

Photo: P. Valentine

24

Bright Cerulean
Jamides aleuas

On the wing this species (**left**), especially the female, can be mistaken for one of the green-banded blues, but the upper surface of the male is a lovely pale blue. Close inspection reveals the underside lacks the metallic green of *Danis* species (see pages 26 and 64). The flight of the Bright Cerulean is usually slow and fluttering with frequent stops at flowers or on twigs. Found from Iron Range through to Wallaman Falls near Ingham this small lycaenid butterfly is at times common at Crystal Cascades, near Cairns. Lycaenidae

Black and White Flat
Tagiades japetus

This is a well named species for it usually displays its distinctive black and white livery with its wings out flat when at rest as shown in the photograph (**opposite above**) but appears as a flashing black and silver blur when in flight. In recent years it appears to have extended its range southwards and has now been recorded at Yeppoon, Magnetic Island and Townsville. Its larval food plant is Common Yam vine, *Dioscorea transversa*, and the caterpillars construct shelters by folding over a leaf, and silking it down to form a low tent. It can be common in the northern part of its range and also occurs in Papua New Guinea with other subspecies as far away as India. Hesperiidae

Theon Jewel
Hypochrysops theon

What a magnificent species this is (**left**), with its brilliant metallic green banding. Confined to the lowland rainforests of the far northern areas of Cape York peninsula it has been seen in the wild by very few people. As recently as 1986 a new subspecies was described from the Rocky River area. The one illustrated is from Iron Range. Males are a pale shining blue above while females are black with cream central areas. The larvae feed on the rhyzomes and fronds of a fern *Drynaria quercifolia*, which grows within the rainforest. Lycaenidae

Banded Demon
Notocrypta waigensis

Wherever wild ginger occurs in northeastern Queensland rainforests a Banded Demon (**opposite below**) may be seen, for larvae feed on the ginger leaves. The caterpillar cuts two small nicks into a leaf about two body lengths apart and curls the section over securing it by silk lines. Adults are particularly fond of nectar and are easily approached when engaged in drinking from a flower. In flight they emit a distinct buzzing produced by the rapidity with which they beat their wings. In the photograph a patch of the white band, which contrasts with the black upper wing, can be seen. This is one of the most common butterflies at Iron Range but also occurs in numbers at Crystal Cascades near Cairns, at Kuranda and in Palmerston National Park near Innisfail. Hesperiidae

Photo: P. Valentine

25

White Nymph
Mynes geoffroyi

In tropical rainforest, especially along streams or in canopy gap areas, the White Nymph butterfly (**right**) may be discovered, usually amongst more densely vegetated regrowth areas. Despite its colourful underside it is difficult to see except when it flies, revealing the white upperwings. Visitors to tropical rainforest in Queensland are usually made aware of the extremely painful results of contact with a Stinging Bush (*Dendrocnide moroides*). Close inspection of these plants reveals many holes in their large leaves. One of the culprits is in fact the larvae of the White Nymph butterfly which feed gregariously underneath the leaf, and which often pupate in small clusters. Nymphalidae

Blue-banded Eggfly
Hypolimnas alimena

Perched on a fallen tree, or on a broad leaf, or even on the ground, the Blue-banded Eggfly will slowly raise and lower its wings in a characteristic fashion (**opposite above**). Males will make brief investigative flights before returning to their favourite perch. Females, which are much duller above, fly close to the ground looking for the larval food plant — a low growing herb with white flowers (*Pseuderanthemum variabile*). The Australian Leafwing (see page 19) also lays its eggs on this plant and both species may occur together. Nymphalidae

Photo: T. J. Hawkeswood / ANT

Large Green-banded Blue
Danis danis

This gorgeous species is often seen in rainforest and wherever one is seen there will usually be others nearby. The brilliant metallic underside (**right**) is in sharp contrast with the upper surface which is white with a pale blue surround and a black margin. About the size of a one dollar coin, the adults are relatively slow-flying compared with many others in the family Lycaenidae (the blues). The yellowish larvae feed on a vine with red or pink juvenile leaves. In common with many other species the larvae only eat the fresh young foliage. They grow very quickly from egg to pupa. Lycaenidae

Red Lacewing
Cethosia cydippe

At Clump Point near Mission Beach these magnificent butterflies are a common sight as they fly briskly along the forest edge or perch with bright red upperwings exposed (**page 28**). Found from Cape York to Townsville it is easiest seen along tracks through rainforest. Even the underside is colourful in this large butterfly (**opposite below**). The female lays her eggs in batches on a fine tendril of the larval food plant, a kind of native passionvine with brilliant red fruit (*Adenia heterophylla*). The gregarious larvae are beautifully banded in black and yellow. Nymphalidae

Red Lacewin

Rainforest Edge & Vine Thicket

Caper White

Caper White
Anaphaeis java

Although the Caper White butterfly (**page 29**) is found throughout Australia, in the tropics it is commonly seen in vine thickets. The photo shows both the male (**above**) and the female. Eggs (**right**) are laid in clusters on a leaf of the larval food plants which are all members of the genus *Capparis* — the caper bushes. These plants are a very common component of vine thickets but they also occur throughout the savannah woodland of tropical Australia. The adults are migratory and at times vast numbers may be seen flying together. Pieridae

Australian Rustic
Cupha prosope

Along the edge of most tropical streams or tracks through rainforest the Australian Rustic (**opposite above**) is a common sight. The dark brown upper wing with its bright orange band makes it easily recognized despite its resemblance to the Lurcher, a much larger butterfly (see page 15). Rustics have an alternating fly-glide flight and usually stay low on the forest edge where their larval food plant occurs. The most commonly used plant is *Scolopia braunii*, a spreading shrub with pinkish-brown fresh shoots which the caterpillars feed upon. The pupa is a beautiful green colour with four pairs of slender spines rising from silver bases. It is suspended below a mature leaf. Other subspecies occur in Papua New Guinea. Nymphalidae

Northern Ringlet
Hypocysta irius

This small brown butterfly is often difficult to find and when discovered will usually be flying close to the ground amongst grasses and shrubs on the edge of rainforest. It also occurs in vine thickets. The upper surface of the forewing has an orange patch (**right**) and underneath it has two eye spots ringed with silver. The larvae of ringlets all feed on grass and are green or brown with horned heads. They are members of the subfamily Satyrinae, known as satyrs or browns, a mainly southern group in Australia. Nymphalidae

Common Aeroplane
Phaedyma shepherdi

There are two Common Aeroplane subspecies in Australia and the photograph (**opposite below**) shows the more northern one, found from Torres Strait islands south to near Townsville. Like the Black and White and the Orange Aeroplanes (pages 16 & 19) this species has a distinctive gliding flight. Common along coastal creeks with fringes of rainforest it also occurs in isolated pockets of vine thicket such as Forty Mile Scrub National Park. Nymphalidae

Red-bodied Swallowtail
Pachliopta polydorus
(Atrophaneura polydorus)

Found from the Torres Strait to Townsville this attractive butterfly (**right**) occurs mainly on the edge of rainforest. At Iron Range it is probably the commonest species flying along the tracks or near the edges of the forest patches. The larvae feed on species of the vine *Aristolochia*, plants also used by Birdwing Butterflies and by the Big Greasy. The egg is laid singly on a leaf and the young larva eats the eggshell before feeding on the plant. The caterpillar is brown with scattered fleshy tubercles of red or yellowish colour. The pupa is very distinctly lobed, quite unlike any other Australian species. They usually pupate under a leaf but have been found beneath rocks amongst vine thicket on Mt White, at Coen. Although commonest in the northern part of their range I have found them on a backyard vine in Townsville and adults have been seen on Magnetic Island. Papilionidae

Common Albatross
Appias paulina

Found from the northwest of Australia across the tropics this species is a strong flyer and although rarely encountered in southern Australia was once found in Launceston. Males and females are differently marked with the female having more extensive areas of black markings above and below. The male (**right**) has a plain yellow underside to the hind wing where the female has a wide dark brown margin. In northern Queensland large numbers of them are sometimes seen along road sides as at Paluma most years. In January 1988 very many were present in vine thickets on Cape York peninsula, laying eggs on their food plant, a densely foliaged shrub. The larva usually lies along the midrib of a leaf when not feeding and its green and yellow colour makes it difficult to see. There are five other albatross species known from the Australian tropics but most are rare (see also page 9). Pieridae

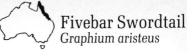

Fivebar Swordtail
Graphium aristeus

The upper surface of this butterfly (**left**) has five distinct bars on the forewing and it is superficially very similar to the Fourbar Swordtail (**below**). In fact they are not closely related and the similarity is more likely explained as convergent evolution. Little was known of this species until January 1988 when the larval stages were discovered in deciduous vine thicket on Cape York peninsula. Few adults had been seen and these only at scattered locations from Rockhampton to Cape York. It appears that this very interesting species flies only in summer and that pupae go into diapause for the rest of the year (arrested development, a kind of hibernation). The pupa may be on the food plant or under rocks at its base. It is believed that the adults emerge from pupae in response to major summer rain and at this time may be locally abundant. Males are known to hilltop, are very aggressive in defending territory and in general fly much more rapidly than Fourbar Swordtails. Green Tree Ants prey upon the larvae and are undoubtedly responsible for significant population control. In recent years adults have been sighted near the beaches north of Cairns and also on hilltops west of Atherton Tableland. All the locations known are in proximity to deciduous vine thickets.

This species also occurs throughout South-east Asia and in Burma and India. Papilionidae

Fourbar Swordtail
Protographium leosthenes

Although more commonly encountered over its range than the Fivebar Swordtail, this species (**left**) is also associated with vine thickets. The adults are commonly seen hilltopping, where they usually glide amongst the shrubs with tails evident behind them. If disturbed they fly quickly away but frequently return to resume their leisurely patrol. The underside of the Fourbar Swordtail lacks the scarlet markings seen in the Fivebar (**above**). Both share the long tails seen in only one other Australian butterfly, the Australian Plane (see page 60). Fourbar Swordtails are also mainly summer butterflies and they can occasionally occur in large numbers. Some summers they are abundant in Forty Mile Scrub National Park and most years they occur on hilltops west of the Atherton Tableland after thunderstorms. The female lays a single egg on a leaf of the larval food plants — members of the family Annonaceae — and the larvae grow rapidly and pupate on the food plant. Adults occur every summer on Mt Stuart, in Townsville. Papilionidae

Photo: P. Valentine

Australian Beak
Libythea geoffroy

This fascinating butterfly is the sole representative of its family in Australia and one of only 10 species in the world. In Africa and America they are called snouts while in Europe and Australia they are beaks. These names refer to the peculiar beak-like projection of labial palpi, evident in the photograph of a female (**left**). Two subspecies occur in Australia with one in the west and another from Cape York to Yeppoon. This species is mainly confined to vine thickets where larval food plants (*Celtis* species) occur. Males are strongly territorial and select preferred perches, usually dead twigs, which they use to rest between vigorous forays after passing butterflies of all species. If a particular male is removed from its perch, within minutes another may occupy it. Females spend much time flying around the food plant and up to two or three may spiral around and within a single tree, seeking sites to lay their eggs. The upperside of the male has beautiful blue forewings while the female upperwings are brown with white patches. Both are similar beneath. Libytheidae

Australian Gull
Cepora perimale

Gulls usually fly with other whites and are particularly common in vine thickets in the wet season when their larval food plants, (*Capparis* species), are covered in fresh growth. The colour of the hindwing underside varies considerably from yellow (**opposite above**) to dark brown. The adults have white upperwings with black margins. The egg is laid singly on the tip of a fresh leaf and the larva is very well camouflaged. The angular pupa is green and attached to the underside of a leaf on the food plant. Pieridae

Orange Lacewing
Cethosia penthesilea

The upper surface of this Northern Territory species (**left**) is more orange than red but it is otherwise very similar to the Red Lacewing found in northern Queensland (see pages 26–28). Larvae feed on the same plant but are quite differently coloured. In recent years captive populations have been established in butterfly farms in Queensland and appear to survive and reproduce very well. It is one of only a handful of species confined to the Northern Territory. Nymphalidae

Canopus Butterfly
Papilio canopus

A large and showy butterfly from northwestern Australia and the Northern Territory, it is also found in Timor. This attractive species (**opposite below**) flies in Darwin gardens where it lays eggs upon introduced citrus trees. The pupa is usually bright green but there is a pale brown form. Sometimes the pupa will enter diapause for as long as two years. It is a close relative of *Papilio fuscus* (see page 16). Papilionidae

Common Banded Awl
Hasora chromus

Awls are a mainly tropical group of skippers in the subfamily Coeliadinae and are medium sized with strong flying ability. This species (**right**) is typical of the group. Occasionally an adult will settle on flowers along the edge of a rainforest track but usually very briefly. Flight is direct and rapid allowing only a few seconds to identify the species. Although few adults are seen together the larvae can be numerous. In Townsville they feed on the fresh foliage of *Pongamia pinnata* and a tree 5–8 metres high may have 100 larvae on it if flushed with new growth. The attractive pale larvae have black and yellow markings and construct shelters by silking a leaf over. They pupate inside the shelter and the pupae are coated in a white powdery substance. As is usual for butterfly larvae which feed exclusively on fresh growth, the period from egg to adult is remarkably short, as little as three weeks in summer months. Other members of this group include the beautifully marked Peacock Awl of northern Queensland and the migratory Brown Awl (see page 46). Hesperiidae

Symmomus Skipper
Trapezites symmomus

The largest member of the Australian Trapezitinae subfamily of skippers, this species is found right down the east coast although divided into three subspecies. The photograph (**right**) shows the northern-most subspecies which is found from Townsville north, although those from Eungella National Park near Mackay appear similar. The upper surface is brown-black with spots and bands of yellow — a handsome species. Adults are rapid flyers and occur wherever their larval food plant is found, clumps of Mat Rush grass *Lomandra longifolia*, now used widely in gardens. In northern Queensland this is usually along the margin of upland tropical rainforest. Eggs are laid singly near the base of the plant and the larvae cut neat triangular pieces from the flat leaves. Later they silk several leaves together to form a tube-like shelter in which they rest during the daytime. Hesperiidae

Savannah Woodland

Northern Jezabel

Grass Jewel
Freyeria trochylus

Perhaps the smallest Australian butterfly, the Grass Jewel (**right**) occurs throughout the savannah woodlands where it flies close to the ground amongst grass and legumes. The row of black spots on the hindwing with metallic capping distinguishes it from other grass blues. The female lays eggs singly on small legumes in the genus *Indigofera* and in Townsville adults are extremely common at times. This is one of the few Australian butterflies also found in Europe, although it is confined there to Greece, Crete and Turkey. It is also found in tropical and subtropical Africa and Asia. Lycaenidae

Northern Jezabel
Delias argenthona

Confined mainly to tropical Australia and Papua New Guinea, this species (**page 37**) is widespread in savannah woodlands. The beautiful scarlet markings underneath stand out in contrast to the white upper surface. The female has more yellow and even the upper surface is tinged with yellow. Eggs are laid in clusters on a fresh leaf of mistletoe and the larvae feed together along the edges of the leaves. It is common for several pupae to occur together under a leaf but invariably most larvae wander away from the mistletoe and may even leave the eucalypt on which it occurs to pupate elsewhere. They fly throughout the year, but there is a winter and a summer form. There are two subspecies in Australia. Pieridae

Small Dingy Skipper
Hesperilla crypsigramma

Throughout its range the best place to see these small skippers is on a hilltop. Along with other species male Small Dingy Skippers (**right**) congregate on the summits of hills where they defend favourite perches from others. This behaviour forms part of a mate-locating strategy whereby females which have freshly emerged from their pupae fly to the top of the nearest hill to find males. Searching is thereby minimised and the female can rapidly begin the essential task of finding larval food plants and laying eggs. Amongst the many skippers which adopt this practice some prefer slightly different parts of the summit environment. The Small Dingy Skipper seems to prefer bare rock areas or bare earth on which to alight. Others prefer dead twigs while some use only leaves on living trees. Females spend only brief periods on a hilltop and are rarely seen by observers compared with males which will spend most of their lives on a single hilltop. If a male is taken from a perch another will usually take his place within minutes. As many as six or seven can be removed with each replacement a little more tatty — presumably older and weaker. The larvae of this species feed on sedges and construct tube-like shelters by silking together several leaves. They pupate within these shelters. There are four other similar species of small dark skippers which also occur in this habitat. Hesperiidae

Common Grass Yellow
Eurema hecabe

Most people in tropical Australia will be familiar with this lovely yellow butterfly (**left**). Of small to medium size it occurs throughout the tropics and is common even in house yards. One of six Australian species it is the most widespread and can be easily recognized from the extent and pattern of black markings on the forewing above. The underside markings vary considerably from almost completely yellow to large patches and mottles of dark brown. The female lays its eggs singly on the leaves of its larval food plants which include numerous herbs or shrubs in the families Euphorbiaceae, Fabaceae and Mimosaceae. The larvae are very well disguised and may rest along the leaf virtually undetectable. The narrow pupa is usually attached by a silk girdle to the underside of a leaf or stem. At times adults can be very numerous and congregate at moist patches of sand to drink. In such a situation on the Normanby River in Lakefield National Park I once saw many Grass Yellow butterflies adhered to the sticky filaments of a sundew (an insectivorous plant in the genus *Drosera*). From a distance the bright yellow patches appeared to be flowers. Several other species were also similarly snared by these plants. Pieridae

Dingy Swallowtail
Papilio anactus

The smallest Australian swallowtail, this species (**left**) is widespread and has probably expanded its distribution by the use of cultivated citrus trees for its larval food plant. It is now established at Alice Springs and recently numerous adults were seen at Bamaga on Cape York, probably due to citrus tree introductions from southern Australia. Adults have not been reported from elsewhere north of about Kuranda. Dingy Swallowtails are known only from Australia and to residents of tropical Queensland are a familiar sight, flying leisurely around gardens in search of nectar, or citrus trees upon which to lay their eggs. The caterpillar is distinct with a bluish-black body with rows of orange-yellow dots. Pupae are narrow and are normally attached to a stem or branchlet of the food plant and are greenish-grey with two projections. This *Papilio* is different from other Australian species and has recently been placed in a subgenus *Eleppone* with close relatives in India and South America. Papilionidae

Miskin's Blue
Theclinesthes miskini

Almost anywhere in Australia this is a common species with adults found throughout open forests and woodlands. They occur on every hilltop in savannah woodlands and their flashing underside draws attention to the incessant aerial combats between males. Every now and then two or more males will fly almost vertically at great speed spiralling to heights of twenty or thirty metres above the canopy of their perching tree. The photograph shows the underside (**right**) with the tail and false eyes of the hindwing. This marking helps confuse predators who snap at the wrong end, thus allowing the butterfly to escape and another chance. It is common to find butterflies with this part of the hindwing removed.

Miskin's Blue is lilac or bluish above in the male while the female is grey-brown with a central patch of pale blue. The egg is laid on many different species of plants and this is one of a handful of butterflies known to use eucalypt leaves. The northeastern Queensland subspecies commonly selects eucalypts as larval food plants but elsewhere mainly wattles are used. The larvae are variable in colour ranging from reddish purple through to green and are usually attended by a few ants. The species of ant in attendance is not constant and there is one report of green ants attending the larvae. At Mareeba I have found numerous larvae on eucalypt leaves attended closely by ants which construct byres at the base of the eucalypt regrowth. In these circumstances the larvae and pupae may be found under the ground within the ants' nest. Lycaenidae

Common White-spot Skipper
Trapezites petalia

In northern Queensland this skipper (**right**) is commonly seen in the savannah woodland areas along the coast. Males sometimes fly on hilltops in company with other species of skippers. Eight of the twelve recognized species of *Trapezites* are found in the tropics but only two are confined there. The larvae of all twelve feed on the strap-like leaves of species of Mat Rush grass (*Lomandra* species) and make characteristic notches when feeding. In most species pupation occurs within the tube-like shelter the larva constructs, however in some the mature larva leaves the food plant and constructs a shelter with leaves and debris some distance from the *Lomandra*.

The Common White-spot Skipper prefers *Lomandra multiflora* as its larval food plant, a species with remarkably variable leaves from blue to green. The female usually lays a single egg near the base of a clump, often on dead leaves or secreted within two overlapping leaves. The larvae are pale green with black heads and usually hide during the day, emerging at night to feed. Hesperiidae

Orange Ringlet
Hypocysta adiante

The upper surface of this species (**left**) is orange with narrow dark margins. As is typical of ringlets, the adults fly slowly close to the ground, in and out of shrubs and clumps of grass. This is the brightest of the five tropical ringlets and the most widely distributed. It is confined to Australia as are all but one member of the genus. A small to medium sized butterfly it is in the Satyrinae subfamily. In common with other ringlets the female lays its eggs on grasses and the larva is slender, pale and with a horned head. Eggs laid in April in northern Queensland complete their cycle to adults in 37 days. The Orange Ringlet is on the wing throughout the year in the tropics. One species of ringlet is confined to southeastern Australia and another to the Iron Range and Coen areas of Cape York peninsula (the Black and White Ringlet) where it is confined to rainforest or vine thickets. The Northern Ringlet is described on page 30. Nymphalidae

Fiery Jewel
Hypochrysops ignitus

One of the gems of the butterfly world, this beautiful species is widespread throughout the more coastal parts of Australia. In northeastern Queensland males occur commonly on hilltops where they usually perch on a leaf of a gum or wattle tree and from where they challenge other butterflies flying near. Their flight is rapid and direct and they very soon return to their perch. Fiery Jewels (**left**) occur in colonies which require two essential ingredients — one of a few species of ants must be present, together with an appropriate plant species. The ants build nests around the base of the food plants — usually regrowth or young plants, often wattles but also Cocky Apple trees and Red Ash. A female will usually lay a small cluster of eggs and the larvae will remain together during their life. Some trees carry large populations of larvae and the effects of their feeding produces dead foliage — the trees take on a burnt appearance. Larvae shelter below the ground in the ant byres or sometimes in the abandoned cocoons of large moths or even in curled leaves. They are always attended by numerous ants. The upperwing of the adult is purple in the male and pale blue in the female. In common with other jewels, the underside is most spectacular. Lycaenidae

41

Large Ant-blue
Acrodipsas brisbanensis

There are seven species of ant-blue butterflies so far known and more are certain to be discovered. They are small but remarkable creatures indeed and are confined entirely to Australia. The larval stages are known from three of the species and it appears that they are carnivores preying upon the larvae of the ants with whom they live. It is thought that the females lay their eggs in clusters at the site of an ant nest. The early larvae may be carried by the ants and even fed by them but at some stage the butterfly caterpillars start to prey upon the ant larvae. Until recently the Large Ant-blue (**right**) was considered rare and confined to a few locations along the east coast from Brisbane south. In the last few years adults have been found in northern Queensland on hilltops west of Paluma and also west of the Atherton Tableland. The photograph is of a male from near Dimbulah in northern Queensland. Others which appear to be this species have turned up on a hilltop near Perth in Western Australia. Males are dark brown above while females have central areas of pale blue. Despite its common name it is a tiny butterfly. Lycaenidae

Photo: P. Valentine

Tailed Emperor
Polyura pyrrhus

This large and magnificent species (**opposite**) is common throughout the tropics and down the east coast of Australia. A powerful flyer, adults are often seen flying rapidly around hilltops. The upper surface is cream and the hindwings have two tails each, as shown in the photograph. The larva is spectacular with a large horned head (**page 44**). Eggs are laid on wattles mainly and the smooth stout pupa is suspended beneath a leaf. I once saw a pupa at Kuranda attached to a leaf right at the entrance to a hive of bees — probably a safe location. The Tailed Emperor is one of only two Australian species in the subfamily Charaxinae, a group of mainly African and Asian butterflies. Adults are often attracted to rotting fruit or sapflows on various kinds of trees. Nymphalidae

Dodd's Azure
Ogyris iphis

The azures are amongst the most beautiful butterflies in Australia with their upper surfaces usually a distinctive shade of blue or purple. Dodd's Azure (**right**) is a brilliant pale blue above and the female has a touch of yellow on the tip of the forewing. In common with other azures the larvae feed upon the foliage of mistletoes and are attended by ants. In northeast Queensland it occurs from the Palmer River on Cape York peninsula to the Burra Range southwest of Townsville. Another subspecies is known from near Darwin. On hilltops west of Paluma the species is common and adults fly along a regular path between outcrops of rock and trees. In flight the flash of pale blue is spectacular but when perched the advantage of the cryptic underwing appearance becomes obvious. Lycaenidae

Tailed Emperor caterpillar (see p 42

Small Dusky Blue
Candalides erinus

As can be seen from the map this species is distributed right across the tropics. Small Dusky Blues (**left**) are often seen feeding at flowers but more commonly flying around the larval food plant, a wiry vine known as Dodder Laurel. Wherever a large patch of dodder occurs so too very often does this butterfly. Eggs are laid on the vine and larvae feed on buds and young shoots. They are very difficult to find. Ten of the twelve species of *Candalides* known in Australia occur in the tropics with one being restricted to the Northern Territory. The upper surface is bronze coloured with a purplish tint in the male while the female is dark brown. The underwings of this group of butterflies are white or grey as in this species. Lycaenidae

Blue Argus
Junonia orithya

Male Blue Argus butterflies (**opposite below**) set up territories usually on patches of bare ground at regular intervals throughout savannah woodlands. Since humans have provided bare patches in the form of roads they can be commonly encountered by walking along a dirt track. They alight directly on the ground and are easily disturbed, however they are not fast enough for some motor cars and frequently they can be found trapped against radiators at the end of a journey through bush roads. Their beautiful blue upper surface is eye catching and the false eyes stand out particularly in the female, which is a duller blue. The underwing is more cryptic but not usually exposed as the butterfly normally spreads its wings out almost flat. Nymphalidae

Chalk White
Elodina parthia

Four species of *Elodina* are currently recognized in Australia but more are sure to be described eventually. All are relatively small for whites and are commonest in the tropics where the larval food plants, Caper Bushes, are often abundant. The upper surface of the Chalk White (**left**) has a dull chalky appearance with a small black patch on the tip of the forewing. Others have more pearly white upper sides and differ in the kind and extent of underwing marking. The eggs are laid singly on a fresh shoot of a Caper Bush and the larvae are particularly difficult to locate as they are so well camouflaged. They usually pupate on the plant. At times the adults are abundant and may be flying with Caper Whites and Australian Gulls (see pages 30 & 35). Pieridae

45

Dark Grass-blue
Zizeeria karsandra

Widely distributed from Sicily and Crete through parts of northern Africa and tropical Asia to Australia this diminutive species (**right**) occurs throughout tropical Australia. It is usually quite local with adults flying commonly around clumps of the larval food plant which is Caltrop, a well known burr legume. This low growing plant has yellow flowers and eggs are laid on the buds which larvae feed upon. Later the tiny green caterpillars feed on leaves often making distinct feeding marks. The spotted underside of the adults help distinguish it from other small blues and the upper surface is dull brown with a central area of pale purple. Lycaenidae

Brown Awl
Badamia exclamationis

Every summer thousands, perhaps millions, of these skippers (**opposite above**) head south on migratory flights from their wintering areas in northeastern Queensland. The larval food plants include Yellow Wood (*Terminalia oblongata*) and this occurs abundantly in the Rockhampton area. Later in the summer the next generation fly northwards in another mass migration. Dull and unexciting to look at this migratory behaviour is remarkable amongst skippers. Hesperiidae

Xanthomera Skipper
Neohesperilla xanthomera

This small orange skipper (**right**) is the most widely distributed member of its endemic Australian genus, being also found outside the tropics in NSW. Despite the importance of this group as Australian butterflies nothing was known of the juvenile stages of any of the four species until 1987. Males of three of the species occur on hilltops and the Xanthomera Skipper is best known due to its presence on numerous hilltops. So far the life history of this species is unknown. The two species of which larvae are known use grasses for food plants and the larvae construct shelters. One of the species (*Neohesperilla senta*) exploits a narrow window in the post-fire succession and this clearly has implications for its survival. In common with other skippers the adults are rapid flyers but frequently settle on twigs, leaves, grass heads or bare ground depending on the species. Hesperiidae

Meadow Argus
Junonia villida

This medium sized attractive butterfly (**opposite below**) has similar habits to its close relative the Blue Argus (page 45) but has a more extensive range in Australia being found even in Tasmania. It is common on other southwestern Pacific islands as well as Papua New Guinea. The Meadow Argus usually flies swiftly close to the ground but frequently settles with wings flat, exposing the beautiful false eyes. Eggs are laid on small herbs including plantains and the caterpillar is spined ("hairy"). Nymphalidae

Pseudictinus Blue
Jalmenus pseudictinus

This finely marked species (**right**) is one of the more recently described Australian butterflies, first recognised in 1967. The adults are very similar to the Ictinus Blue (*Jalmenus ictinus*) and the range of each overlaps through central and southern Queensland and it is virtually impossible to separate them even when held in the hand. The feature which makes them clearly different species is their distinctly different juvenile stages and the two completely different ant species which attend the larvae. The Pseudinctinus Blue larvae are attended by numerous small black and tan ants (*Froggatella kirbyi*) and in northern Queensland feed upon *Acacia flavescens*. They occur in local colonies often numbering many hundreds of larvae on a few trees. Lycaenidae

Common Brown Ringlet
Hypocysta metirius

Typical of ringlets the flight of this medium small butterfly (**below**) is usually close to the ground weaving amongst shrubs and grasses. Very like the Northern Ringlet (see page 30) it lacks the orange forewing of that species. The undersides of both are similar in having a pair of "eyes' ringed in silver beneath each hindwing, similar to that shown for the Orange Ringlet (see page 41). The pale green egg is laid on grasses and the larva is slender with a horned head. It is in the subfamily Satyrinae. Nymphalidae

Big Greasy Butterfly
Cressida cressida

What an unfortunate name for a species which is really very attractive! It is no doubt derived from older female specimens which have lost most of their scales and have wings which appear very much like old-fashioned grease-proof paper. In this species the male (**left**) is much more colourful and offers no clue to the derivation of the common name. After mating the male deposits a sphragis beneath the female's abdomen which covers her genitalia — a kind of butterfly chastity belt. It is one of the commonest butterflies of the tropics with different subspecies in the Northern Territory and eastern Australia. The closest relatives of this species live in Argentina and thereby pose a puzzle for biogeographers. Papilionidae

Glasswing
Acraea andromacha

Africa is the main home for the subfamily Acraeinae and in Australia the single representative is the Glasswing (**below**). Both sexes are similar and have almost transparent wings. Females lay their eggs in clusters on native passion vines and the larvae feed together and may even pupate near one another. Like the Big Greasy (**left**), Glasswing males deposit a "chastity belt' (sphragis) over the female's genitalia after mating. This is thought to inhibit additional copulation with the female by other males. Nymphalidae

oto: P. Valentine

Northern Imperial Blue
Jalmenus eichhorni

This species is confined to Cape York peninsula, as far south as Forty Mile Scrub National Park. It occurs in local populations usually centred on a group of *Acacia* plants upon which larvae feed. The photograph (**left**) shows the distinctive underside markings and the fine tails to the hindwing. The upperside has a central blue-green area in the male with a pale blue central zone in the female. Nine species are so far described in the endemic Australian genus *Jalmenus* with the probability of at least one more from Western Australia. These butterflies have fascinating relationships with ants and have been the subject of considerable study. The larvae of the Northern Imperial Blue are attended by the large red Mound Ant which collects nectar from the caterpillars. In return the ants reduce the rate of parasitism and appear to inhibit predators. Recent studies indicate a true mutualism as both ants and butterflies benefit from the association. Lycaenidae

Purple Azure
Ogyris zosine

Widely distributed throughout the Australian tropics this large blue is quite beautiful above but very cryptic below (**opposite above**). The underside makes the butterfly very difficult to see when it alights on the trunk of a gum tree or a dead twig. In the photograph a nick in the hindwing barely exposes some of the yellow patch on the forewing which identifies a female. The male upperwing is a beautiful purple but in the female it may be purple, purplish-blue or bright blue. In northern Queensland the colour of the female seems to reflect relative humidity; around Townsville they are mainly blue while they are mostly purple at Cairns. The upper surface colour is only evident in flight. The larvae feed upon mistletoe leaves and are attended by sugar ants. Lycaenidae

Common Grass-blue
Zizina labradus

The Common Grass-blue is widespread in savannah woodland in northern Australia. The photograph (**left**) shows a mating pair but hides the upperside which is blue in the male and brown with a central patch of blue in the female. Adults usually fly close to the ground in a slow meandering pattern. They are small butterflies and the larvae feed upon legumes, including clovers, in urban lawns. Lycaenidae

Dingy Ring
Xois arctoa

An inconspicuous species the Dingy Ring is readily identified from the eyes evident on the underside (**opposite below**). In flight it is slow and appears clumsy although when startled may rapidly disappear. It usually flies close to the ground amongst the clumps of grass typical of its preferred habitat. The green egg is laid on grass species and the larvae, which may be pale brown or green, are distinctly hairy. Nymphalidae

51

Coast & Wetlands

Lesser Wande

Lesser Wanderer
Danaus chrysippus

Although found in all habitats in the tropics this species is frequently common in coastal areas, especially near mangroves. The Lesser Wanderer (**opposite**) is one of the milkweed butterflies (subfamily Danainae) which includes many species of crows and tigers. The larval food plants of these butterflies have a milky sap and are toxic which confers some protection on the adult butterflies. The female of one of the nymph butterflies (the Danaid Eggfly) mimics this species although the male is completely different in colour. One fascinating aspect of the Lesser Wanderer is that it is one of the first species of butterflies on record, depicted some 3500 years ago on an Egyptian tomb. The larvae are colourful and have six fleshy filaments. Nymphalidae

Narcissus Jewel
Hypochrysops narcissus

Named for the Greek God infatuated with his own reflection, this is indeed a beautiful species (**left**). The upper surface is a rich dark blue which flashes in the sunlight when the butterfly is in flight. However it is the underside which is evident at rest and that is richly marked in red and white and green fleckings. The best place to see these lovely creatures is in mangroves where they may be seen "dogfighting" above the canopy. Larvae feed on the leaves of at least two species of mangroves as well as a mistletoe. They also occur in swampy areas and in rainforest at Iron Range and can be particularly common at Yule Point north of Cairns. Lycaenidae

Copper Jewel
Hypochrysops apelles

The upper surface of this lovely jewel (**front cover, lower left**) is bright copper but this is hidden except when it is in flight. In Townsville the Copper Jewel is commonly seen flying around Cocky Apple trees, one of its many larval food plants. The caterpillars are usually solitary and are attended by small black ants with triangular abdomens. The more usual habitat for this species is mangroves. Once at Cape Cleveland, near Townsville, I saw hundreds of adults within a distance of two hundred metres along the edge of a mangrove community. They had recently emerged and were busy disputing territory or laying eggs. Lycaenidae

Common Swift
Pelopidas agna

Swampy areas or stream banks with thick clumps of grass are favoured by this rapid-flying skipper (**left**). The female lays an egg singly on a blade of grass and the larva builds a shelter by silking two sides together. Later the pupa is suspended beneath a blade of grass and matches the green colour very well. This species, and other Australian swifts, may prove troublesome for the developing rice industry as they sometimes lay their eggs on young rice plants. Hesperiidae

Eichhorn's Crow
Euploea eichhorni
(alcathoe)

In the dry season many of these butterflies (**right**) congregate along the many streams and rivers of Cape York peninsula. They appear similar to the Common Crow (pages 68 & 71) but lack the extent of marking on either upper or under sides and the males lack a sexmark. The upper side of Eichhorn's Crow is a rich brown-black with a small cluster of white marks near the tip of the forewing. The close similarity of Australian crows is thought to confer mutually reinforcing protection from predators for all are likely to be toxic to potential predators due to the plants eaten by their larvae. Nymphalidae

Black and White Tiger
Danaus affinis

Along the coastal parts of tropical Australia this attractive species is usually common. The photograph shows a female taking nectar from a flower (**opposite above**). Males have a distinct dark pouch on one of the hind wing veins, called a sex mark. In this group of butterflies (Danainae) males have "hair-pencils" which give off a scent thought to attract females. The larvae mostly feed on plants with milky sap and experiments have shown the adults are unpalatable to birds. Most species are long-lived for butterflies with some living for six months or more. Nymphalidae

Saltpan Blue
Theclinesthes sulpitius

Confined to the saltpan areas of mangroves, tidal flats and estuaries, this small species (**right**) occurs from Cooktown to Victoria. The larvae feed on Samphire and other saltbushes and the photograph shows an adult on one of these food plants, taken in Townsville. The upper side is brown. Adults fly commonly very close to the ground and often in very large numbers around their larval food plants. Lycaenidae

Lower's Darter
Telicota mesoptis

This striking orange coloured skipper (**opposite below**) is found mainly in the wetter parts of the tropics from Cape York to Mackay. It occurs in grassy areas along the edges of streams or swamps or lowland rainforest. There are seven other species of darter in northern Queensland and all are similar, making identification very difficult for the novice. In some cases dissection and careful examination of genitalia seems the only certain way to separate species. In the field there are some differences in preferred habitats which help distinguish one from another. Hesperiidae

Orange Dart
Suniana sunias

This flighty little skipper is widely distributed in the tropics but especially so in wet areas. Shown in a typical pose (**right**) the Orange Dart is common wherever a sward of dense green grass is growing. The males will rest on blades of grass and aggressively defend their patch by flying rapidly after any passing butterfly before returning to their perch. Females seek out a variety of grasses on which to lay their eggs. The long slim larvae are green with a black head, and they construct tubular shelters by rolling a blade of grass. They emerge to feed on other parts of the same blade of grass, eventually leaving nothing but the midrib and their shelter. At this stage they move on to another blade and a new shelter.

In urban areas of the tropics this species is common in gardens wherever grass is left uncut. The Orange Dart is a member of the subfamily Hesperiinae which is a group including many orange and black skippers with superficial similarity. It is often very difficult to tell one species from another. Hesperiidae

Cooktown Azure
Ogyris aenone

The Cooktown Azure (**right**) has a very interesting distribution for it occurs in *Melaleuca* swamps in northern Queensland and also in an isolated pocket on the Darling Downs. In northern Queensland the larvae feed upon mistletoes which occur on *Melaleuca* trees while at Leyburn in the Darling Downs they feed on mistletoe on *Casuarina* trees. So far they have not been discovered in the intervening country between their known northern locations (Cape York to Ayr, near Townsville) and the colony at Leyburn.

As with other azures the underside is highly cryptic, as can be seen in the photograph, and makes the butterfly difficult to see when it alights on trees or mistletoe plants. The upper surface of the Cooktown Azure is an exquisite pale blue colour which catches the sun brilliantly when it flies. A hint of this upper wing colour is present near the edge of the lower hindwing where a few scales are showing. The photograph shows a freshly emerged adult on its mistletoe food plant from near Cardwell in northeastern Queensland. (see also pages 42 & 51) Lycaenidae

Photo: P. Valentin

Tailed Cupid
Everes lacturnus

This attractive little blue (**left**) is common throughout the wet season when it may be seen flying in wetter areas. It is a slow-flying species and adults congregate in areas where the larval food plant occurs. This is a pea shrub (near Ingham *Desmodium heterocarpon*) with dense clusters of flowers followed by pods which occur in large clumps. The larvae feed initially on the flowers and then later eat the green pods. They pupate within the clumps of pods where they are difficult to see. The green larva is attended by ants and the pupa is pale. The underside of this species is distinctive with orange spots not present on other butterflies. The upper side is purplish-blue in the male but brown-black with the orange patches in the female. The tails are particularly long and fine and help distinguish the Tailed Cupid from other species. It occurs from India and China through South-east Asia to Indonesia and Papua New Guinea. Lycaenidae

Orange Bushbrown
Mycalesis terminus

Along virtually any creek or stream through the *Melaleuca* swamps of northeastern Queensland there will be numerous Orange Bushbrowns (**left**). They are also common wherever clumps of blady grass occur, especially along the edge of rainforest. The flight is varied as they may at times undulate amongst grass clumps or weave around tree trunks but on other occasions may fly quite directly and rapidly. This species is attracted to rotting fruit and adults will often sit for long periods sipping juices from fallen mangoes or other soft fruit. Recent studies at James Cook University have demonstrated that adults will lay eggs on many species of grasses. The larva is green or pinkish-brown with a horned head and the green pupa is suspended head downward — usually near the base of the grass clump. There are two other species of bushbrowns in Australia both of which are found in the Northern Territory as well as in northeastern Queensland. The Dingy Bushbrown (*Mycalesis perseus*), often flies with the Orange Bushbrown but is a more or less uniform dark grey colour but with false eyes obvious beneath the hindwing. The third species, the Cedar Bushbrown, is shown on page 59. These are all members of the subfamily Satyrinae (the browns). Nymphalidae

Blue Tiger
Danaus hamatus
(Tirumala hamata)

One of the milkweed butterflies (subfamily Danainae), the Blue Tiger (**right**) is usually common in coastal areas of northern Queensland. During the wet season it disperses quite widely but in winter the adult butterflies come together in large aggregations of many hundreds or even thousands of butterflies, usually along creeks and near the coast. At this time of the year, the dry season, the Blue Tiger does not reproduce and is often seen throughout the day perched on stems and twigs or dead branches beneath the canopy of the forest. The Common Australian Crow also forms over-wintering clusters and a photograph is included of a small part of such an aggregation (see page 71).

Although a specimen of the Blue Tiger was not included in the Banks collection there is a description in his journal of a vast aggregation of butterflies which appear to have been Blue Tigers. On the 29th May 1770, when the Endeavour was about halfway between Rockhampton and Mackay, Banks recorded seeing "acres" of them! This long-lived species is a powerful flyer and one year during a trip to Wheeler Reef off Townsville I counted dozens up to 50 km offshore. It is therefore not surprising to find the species well distributed on Pacific islands including Fiji, Tonga, Samoa and New Zealand. Nymphalidae

Pale Ciliate Blue
Anthene lycaenoides

This little blue butterfly (**right**) is not as common as its close relative the Dark Ciliate Blue (see page 20) but it has many similar habits. Adults frequently remain close to the larval food plant, flying briefly out from the tree and returning to perch. They also spend some time walking over the branches and leaves. In the Townsville area the eggs are usually laid on species of *Cassia* trees, typically on the flower buds and almost always in the presence of small black ants (*Paratrechina bourbonica*). The larvae are dark green with yellow "arrows' on the upper surface. Occasionally the larvae are attended by Green Tree Ants. Unlike those of the Dark Ciliate Blue the pupa is always beneath a leaf and is coloured green which makes it very difficult to see. The Pale Ciliate Blue also lays its eggs on Litchi trees and the larvae eat the buds and flowers. It is quite common in urban areas. Lycaenidae

Eastern Brown Crow
Euploea tulliolis

This striking species is found along the east coast from Cape York to northern NSW. The photograph (**left**) shows the brown hindwings contrasting with the velvet-black forewings, a feature which readily distinguishes this species from other Australian crows. Generally the Eastern Brown Crow flies slowly and low but if disturbed will fly higher and much faster. It does settle frequently, often on dead twigs and sometimes at puddles of water. In common with other milkweed butterflies (subfamily Danainae), it sometimes occurs in large clusters and overwinters along the coast. A good place to see this species is along the margins of coastal streams and rivers or behind mangrove communities. It is not uncommon that several species of danaid butterflies may fly together. A comparison may be made with the Common Australian Crow shown on page 68. As with other crows, the eggs are laid singly on the food plant which is usually the rambling vine *Malaisia scandens*. Nymphalidae

Cedar Bushbrown
Mycalesis sirius

Found in the Northern Territory and Cape York to Mackay the Cedar Bushbrown (**left**) prefers wetter swampy areas compared with the other bushbrowns (see page 57) although in some places all three species fly together. The upper wings are reddish-brown with less conspicuous ocelli ("eyes") compared with the under wing shown in the photograph. The rich cedar colour readily distinguish it from the other two species. In flight this species is relatively slow and meanders around clumps of grass, settling on blades or sometimes on the ground. Eggs are laid singly on blades of grass (usually Guinea Grass, not Blady Grass) and the larvae feed openly on the leaves at night. This species is easily reared in captivity and adults readily mate in small shadehouses. The Cedar Bushbrown is one of only five Australian species of the subfamily Satyrinae confined entirely to the tropics. Nymphalidae

59

Australian Plane
Bindahara phocides

Found from Torres Strait to Townsville this medium-sized species is readily recognized by its impressive tails (**right**). The male, underside shown, is a rich black above with two central patches of dark blue on its hindwings. The female is brown-black above with a large patch of white on each hindwing and with the underwing white with brown markings. The Australian Plane is a rapid flyer, males often setting up territory on the forest canopy. Although they are seen in rainforests, they are better known from the coastal zone, usually right above the high water mark where their larval food plant, the vine *Salacia chinensis*, occurs. At both Cape Tribulation and Mission Beach the females may be commonly observed laying eggs upon the fruits of this plant. The tiny larva bores into the fruit and eats the seeds within. Large larvae leave the fruit to pupate. Lycaenidae

Apollo Jewel
Hypochrysops apollo

In the paperbark (*Melaleuca*) swamps and mangroves between Cape York and Ingham a strange bulbous epiphyte, the Ant-plant, occurs. In places these plants also occur in lowland rainforest but they are essentially coastal, even occuring in the magnificent dunefield vegetation of Shelburne Bay. The Apollo Jewel (**opposite above**) could be as well named the Ant-plant Jewel for it is within the hidden galleries of this plant that the larva grows, feeding upon the fleshy interior and emerging at night to feed upon the bright green leaves. Once it has reached maturity it cuts a small circular exit hole, and then pupates within the Ant-plant. The myriads of tiny golden ants which occur within these plants do not harm the butterfly larvae but instead closely attend them and take a fluid from special glands. This entire relationship is one of the most fascinating examples of mutualism and symbiosis; where plant benefits from ants, ants from plants and butterfly from both. The adult butterfly is particularly exquisite, being a bright orange above. Lycaenidae

Moth Butterfly
Liphyra brassolis

It is perhaps difficult to imagine a butterfly as a carnivore but that is indeed the case with this amazing species (**opposite below**). Female Moth Butterflies lay their eggs on branches not far from brood nests of the aggressive Green Tree Ant and the tiny larvae make their way to the inside of the nest where they spend the rest of their larval days. To inhibit damage by green ants Moth Butterfly larvae have developed a flattened shape and tough skin. They prey upon the ant larvae, chewing them with an impressive set of mandibles. Even at pupation the butterfly protects itself, by pupating within its final larval skin! Upon emerging the adult has numerous small white temporary scales over its body and wings which dislodge as ants attack enabling escape to be made from the nest. Lycaenidae

Photo: P. Sams

Urban

Lemon Migra

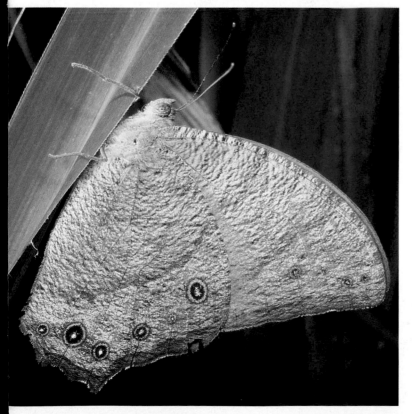

Zebra Blue
Syntarucus plinius

This delightful little blue (**left**) is also called the Plumbago Blue as the larvae feed upon the buds and flowers of *Plumbago zeylanica*, a native species commonly used in gardens due to its beautiful blue flowers. The photograph also shows this plant. In Queens Gardens, Townsville, there is a maze which has been partially constructed from plumbago plants and there are always adults of this butterfly flying around them. The upper surface contrasts with the underside shown and is lilac in the male and brown with pale blue and white in the female. A small clump or two of plumbago in a backyard in northern Queensland guarantees a few of these distinctive butterflies flying most of the year. They have a fluttering flight style and usually settle on the food plant after several tentative approaches. Lycaenidae

Lemon Migrant
Catopsilia pomona

Widely distributed throughout the tropics this swift-flying butterfly (**opposite**) is common in all urban areas. The adults are quite variable and occur in colours ranging from lemon-yellow to cream. One form has pink, another black, antennae. The females have secondary markings of black. One reason for urban proliferation is that the common garden tree *Cassia fistula* is used as a larval food plant. The spindle-shaped egg is laid singly on a young leaf and the green larva usually rests along the midrib. Pupation occurs beneath a leaf or stem and the pupa is green, however after the adult emerges the pupal skin is pale white. Occasionally young trees are stripped of their leaves and in there place may be seen dozens of empty pupal cases festooning the bare branches. Pieridae

Evening Brown
Melanitis leda

With the onset of late afternoon this species becomes active, flying close to the ground amongst shrubs and grass clumps. Found throughout the tropics from Africa to Asia and into the Pacific Ocean islands, the Australian subspecies (**left**) is one of the best known garden butterflies. Its larval food plants are coarse grasses and a clump of uncut Guinea Grass will regularly produce a crop of Evening Brown butterflies. The caterpillar is green with a pair of long black horns. It rests on the underside of a blade of grass and subsequently pupates in a similar location. There are seasonal forms of the Evening Brown and the one shown is the tropical summer form. The winter form has much reduced "eye" spots underneath but darker underwing markings with more prominant upperwing markings. In each form the upperwing is mainly rich brown but the winter form has a large central area of paler brownish-orange. In common with many other members of the subfamily Satyrinae, adults frequently alight on the ground and it is then that the excellent camouflage effect of the underwing pattern may be appreciated. Nymphalidae

Wanderer
Danaus plexippus

Perhaps the world's best known butterfly (**right**), this species is known in North America as the Monarch. There it undertakes an annual two-way, north-south, migration and in winter assembles in enormous aggregations at a few known locations. These winter assemblies have long been significant tourist destinations for the sight is truly remarkable — entire trees clothed with large and colourful butterflies. The Wanderer settled in Australia only after European colonization, when its larval food plants (cotton bushes) were established. Since the colonial era it has also established on numerous Pacific islands and there is still debate over whether dispersal occured naturally or with human assistance. Whatever the answer to that question it appears the Wanderer was established in Australia by late last century and there have been recent reports of clusters of butterflies overwintering in southern Australia. This large orange butterfly is frequently seen in northeastern Australia soaring and swooping or flying rapidly, especially in more open areas and farmlands. Nymphalidae

Common Eggfly
Hypolimnas bolina

In this large and showy species males (**opposite above**) differ sharply from the females (**opposite below**) although the underwings of both are similar. They fly with abrupt undulations and alight frequently at flowers to take nectar. When at rest in sunlight both sexes usually open their wings to display the colourful upper surface. Females vary considerably in their upperwing colours and occasionally the central orange patch is absent. They lay their eggs on the underside of a leaf of the larval food plants which include many common garden weeds and cultivars (e.g. Sida-retusa and Joyweed) and adults fly commonly in gardens. Males establish territories and from a perch on a leaf will venture forth to challenge any passing butterfly or even other flying insects. Nymphalidae

Small Green-banded Blue
Danis hymetus

Although mainly a rainforest butterfly, this tiny blue is a very common visitor to gardens in most parts of northeastern Queensland. It is usually seen feeding at Lantana blossoms or other garden flowers. In the photograph (**right**) the lovely underside is shown with its metallic green colours. The male has a blue upperwing while the female is blue and white above. The larval food plant is a rainforest pioneer (*Alphitonia* species) now grown often in gardens. It is difficult to see the pale white larvae at rest beneath the leaves. Along the edge of roads and tracks in wetter parts of northeastern Queensland seedlings and young regrowth of *Alphitonia* are common and here also the Small Green-banded Blue may be found. But they also occur in less wet areas including urban parts of Townsville. Two other species of *Danis* occur in Australia and these are figured on pages 21 and 26. Lycaenidae

Orchard Butterfly
Papilio aegeus

This showy species is one of our largest butterflies and one of the commonest in eastern Australia. The photograph (**left**) shows a male (right) courting a female (left). Captured in mid-flight the male is attempting to stimulate the female by a distinctive, almost stationary, fluttering action, common to many butterflies. In some species this courting flight is very synchronized and in others it appears clumsy as the female is more or less herded by the male. If the response by the female is favourable then copulation follows and may last for an hour or more. The Orchard Butterfly is so named due to its use of citrus trees as larval food plants and many private gardens with a lime or lemon tree help maintain this species in urban areas. The small larvae resemble bird droppings, a strategy also adopted by other species in this family. Papilionidae

Dull Oakblue
Arhopala centaurus

In coastal northern Queensland this species (**opposite above**) may be seen almost anywhere that one of its larval food plants occurs in the presence of Green Tree Ants. Like the other oakblues (see pages 21 & 22) the Dull Oakblue has a brilliant blue upper surface contrasting with a dull underwing colour. Larvae are attended by Green Tree Ants and the most common food plants are in the genus *Terminalia* which is widespread in the tropics. At Anderson Park in Townsville there are often many hundreds on the wing and this occurs in coastal towns throughout the tropics. Another subspecies occurs in the Northern Territory. Lycaenidae

Pale Green Triangle
Graphium eurypylus

This attractive species (**left**) commonly flies in gardens as adults seek out the introduced soursop tree, grown widely for its fruit, as a larval food plant. Pale Green Triangles may be seen flying around any tree with fresh shoots. If Green Tree Ants have infested the tree the females will not lay eggs for these ants eat the larvae. The Pale Green Triangle has a nervous fluttering flight when it is feeding at flowers similar to that of the other triangle species (see pages 6 & 19). Papilionidae

Yellow Palmdart
Cephrenes trichopepla

If you have ever noticed tatty palms and wondered how they may have got that way — look no further. The most likely explanation is that the Yellow Palmdart (**opposite below**), or its close relative the Orange Palmdart, has laid eggs on the leaves and their larvae have been happily chewing away. Many ornamental palms take on a decidedly untidy appearance when the palmdarts have finished with them. The adult often flies in gardens, frequently perching on clothes hoists and enjoying the sun. They are rapid and powerful flyers. Hesperiidae

Chequered Swallowtail
Papilio demoleus

ound throughout Australia, and indeed from
 through India, China and South-east Asia, it is
mes a very common species. In tropical
tralia the Chequered Swallowtail (**opposite
ve**) seems to occur in waves as a migration
es through in search of larval food plants. On
e occasions they will lay eggs on citrus trees
er than the more usual legumes of inland
tralia (eg Emu-foot). Once a female returned to
garden four days in a row and on each occasion
about 20 eggs on a Pink Evodia tree (*Euodia
yana*) planted for the Ulysses Butterfly. I
sferred the larvae to a lime tree and they
pleted their life cycle. The adults have a direct
very rapid flight, close to the ground. There are
rts of large migrations in the Northern Territory
elsewhere. They can be local, however, as I have
d them at certain places along the coast. On a
y coastline in the Townsville Town Common
ironmental Park they fly amongst the coastal
bs most months of the year. Papilionidae

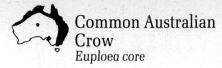

Common Australian Crow
Euploea core

This very familiar species looks much like all the
other crows but has many more white markings
(**opposite below**). It is sometimes called the Oleander
Crow for it frequently lays its eggs on that
introduced plant. Gardens with young plants will
almost certainly have larvae on them from time to
time. The larvae are quite colourful with narrow
black and white bands and four pairs of fleshy
filaments. The pupa is remarkable for it turns a
brilliant shining silver or gold, seemingly metallic,
with a few markings. It is suspended beneath a leaf
and just before the adult emerges becomes black and
white as the adult's colours show through the
transparent pupal skin. The larvae also feed upon
that pest of the tropics the introduced Rubber Vine,
now rampant in many parts of the country. In the
winter months many Common Australian Crows
form large aggregations, usually in sheltered gullies
and creeks and a typical cluster is shown on **page 71**.
This is one of the "milkweed" butterflies in the
subfamily Danainae. Nymphalidae

Further Reading

ACKERY, P.R. & VANE-WRIGHT, R.I.
1984
Milkweed Butterflies
British Museum (N.H.), London

BARRETT,C. & BURNS, A.N.
1951
Butterflies of Australia and New Guinea
Seward, Melbourne

COMMON, I.F.B. & WATERHOUSE, D.F.
1981
Butterflies of Australia
2nd edition, Angus & Robertson, Melbourne

D'ABRERA, B.
1977
Butterflies of the Australian Region
2nd edition, Lansdowne, Melbourne

EMMEL, T.C.
1976
Butterflies
Thames & Hudson, London

McCUBBIN, C.
1971
Australian Butterflies
Nelson, Melbourne

OWEN, D.F.
1971
Tropical Butterflies
Clarendon, Oxford

SMART, P.
1975
The International Butterfly Book
Ure Smith, Sydney

VANE-WRIGHT, R.I. & ACKERY, P.R.
(Editors) 1984
The Biology of Butterflies
Academic Press, London

he Entomological Society of Queensland publishes a journal in which many recent discoveries about Australian
terflies are described. Enquiries may be directed to
stralian Entomological Magazine,
. Box 537, Indooroopilly, Qld. 4058

Printed by Inprint Limited Australia

Index to Butterflies